Houses and Equipment For Laying Hens

by US Dept of Agriculture

with an introduction by Jackson Chambers

This work contains material that was originally published in 1960.

This publication is within the Public Domain and was originally published with Public Funding for the Public Benefit.

This edition is reprinted for educational purposes and in accordance with all applicable Federal Laws.

Introduction Copyright 2017 by Jackson Chambers

Self Reliance Books

Get more historic titles on animal and stock breeding, gardening and old fashioned skills by visiting us at:

http://selfreliancebooks.blogspot.com/

Introduction

I am pleased to present yet another title on Poultry.

The work is in the Public Domain and is re-printed here in accordance with Federal Laws.

As with all reprinted books of this age that are intended to perfectly reproduce the original edition, considerable pains and effort had to be undertaken to correct fading and sometimes outright damage to existing proofs of this title. At times, this task is quite monumental, requiring an almost total "rebuilding" of some pages from digital proofs of multiple copies. Despite this, imperfections still sometimes exist in the final proof and may detract from the visual appearance of the text.

I hope you enjoy reading this book as much as I enjoyed making it available to readers again.

Jackson Chambers

CONTENTS

	Page
Methods of housing	1
Testing effects of housing and flock sizes	1
Influence of climate on laying-house design	2
Four types of laying houses	2
Special problems of the small flock	3
Special problems of the large flock	4
Space requirements	4
Summer comfort	4
Winter comfort	6
Systems of ventilation	6
Insulation and wall finish	8
Vapor barrier	8
Building the laying house	10
Construction costs	10
Location and orientation	11
Selecting a plan	11
Selecting building materials	12

	Page
Building the laying house—Continued	
Building details	12
Footing and foundations	12
Floors	13
Wood framing	13
Roof covering	14
Windows	15
Interior details	16
Nests	16
Feeders for mash, oystershell, and grit	19
Waterers, and piping of water to fountain	20
Roosts, droppings pit, droppings board, and utility pit	21
Feed room	22
Egg-handling room or workroom	24
Egg-holding room or cabinet	26
Electrical outlets, lights, and standby generators	26

ACKNOWLEDGMENTS

The author is indebted to agricultural engineers and poultry specialists at many of the State agricultural colleges and also within the United States Department of Agriculture; to individual poultry farmers; to representatives of poultry magazines and building-materials manufacturers; and to members of the Barnyard and Poultry Equipment Council, for comments, suggestions, and photographs.

This publication supersedes Farmers' Bulletin 1554, Poultry Houses and Fixtures.

Washington, D. C.

November 1956
Slightly revised October 1960

Houses and Equipment
FOR LAYING HENS...
for loose housing

By Hajime Ota, agricultural engineer, Agricultural Engineering Research Division, Agricultural Research Service

Methods of Housing

The right kind of a house for laying hens is extremely important to the poultryman. It facilitates a high rate and quality of egg production, results in economy of feed, and helps insure freedom from disease. It accomplishes these results by providing light, dry, clean, and well-ventilated quarters for the hens; also, by protecting the hens from wind, rain, and snow, as well as from sudden changes in temperature and from extreme heat and cold.

The interior of a laying house should be so arranged and equipped that both the hens and the eggs can be cared for with a minimum of labor and time. Construction and upkeep costs should be reasonable, and the house should present a neat appearance.

This bulletin deals with loose housing of hens. This is one of the two most widely used methods of housing laying hens. The other method is to use wire cages.

When loose housing is used, the hens are kept in a building in which they are free to wander about. The floor is covered with litter (straw, or similar material) or the hens are on wood-slat platforms raised about 12 inches above the floor. The advantages of loose housing over wire cages include:

    Lower investments per laying hen.
    Less brooding.
    Less fly-control difficulty.
    Less labor spent looking after equipment.
    No wire floors to cause condensation of water during foggy weather, resulting in wire-marked eggs.
    Less need for heating buildings in zones 1 and 2 (fig. 1) during cold weather.

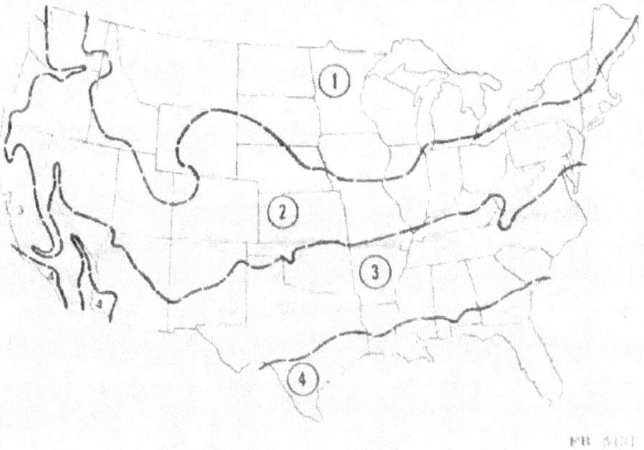

The cage method is well adapted to farm-building zones 3 and 4 (fig. 1). Cages are also used in well-insulated houses in building zones 1 and 2. Wire cages up to 12 inches wide may hold 1, 2, or 3 hens depending on the size of the hens. Wire enclosures more than 12 inches wide with capacity of from 3 to 100 hens are known as colony or wire pens. The advantages include:

    More precise culling.
    Lower feed costs.
    Better control over parasites and disease.
    More uniform egg production.
    Less competition for feed and water.

Testing Effects of Housing and Flock Sizes

The important effect of good housing on production and feed consumption was shown by a series of 3- to 6-week tests at different constant-air temperatures, conducted by the United States Agricultural Research Service, Beltsville, Md. (1951-54). Ten locally hatched Rhode Island Red hens, in their first year of production, were kept on litter in each test. Air temperature, relative humidity, ventilation rate, lighted hours, water, and litter management were controlled.

The investigation showed that as egg production fell off at low temperature, weight per egg increased slightly. Small eggs with poor shells

FIGURE 1.—Farm building-zone map, based on January temperatures.

TABLE 1.—*Egg production and feed consumption of Rhode Island Red hens at various constant temperatures and 75 percent relative humidity (1951-54)* [1]

Air temperature	Eggs per day per 100 hens	Weight of eggs per dozen	Eggs per day	Feed consumption per day per 100 hens	Feed per pound of eggs	Feed per dozen eggs
°F.	Number	Ounces	Pounds	Pounds	Pounds	Pounds
37	65	23.9	8.8	35	4.0	6.5
45	74	23.8	9.1	33	3.5	5.4
55	78	23.5	9.5	31	3.3	4.8
65	75	23.2	8.9	29	3.3	4.6
75	68	22.7	7.9	27	3.4	4.8
85	56	22.1	6.5	25	3.9	5.4

[1] After a 10- to 14-day period of acclimatization, tests were run for 3 to 6 weeks. Ration was the same at all temperatures.

were laid at temperatures above 80° to 85° F. (table 1). The least weight of feed per pound of eggs was required at temperatures between 45° and 65°, the feed requirement at 35° being 40 percent more than at 55°. The ration was the same throughout the series of tests.

The results also showed that protecting the hens from low temperatures saves feed, while protecting them from high temperatures results in higher rate of egg production.

Studies by the United States Department of Agriculture (1950) showed that the average time spent in caring for laying hens in flocks of more than 200 was 1.5 hours of work by 1 man per hen or pullet per year. In flocks of less than 200, the average time spent was 2.5 hours per hen or pullet per year. Other State and Federal studies have shown that the time spent as well as the distance walked, can be reduced with large flocks, improved housing, layout, and mechanization. As low as 0.9 man-hour per hen per year have been reported (1957).

Influence of Climate on Laying-House Design

Differences in climate account for much of the variation in types of poultry houses used in the different States. A poultryman considering a new type of house should compare his own climate with that of the locality where such a house has been used successfully.

Temperature, sunshine, and relative humidity are probably the most important climatic factors to be considered. Wind is also an important factor in determining the structural design of the house.

A farm building-zone map, based on January temperature and relative humidity, is shown (fig. 1). Hours per day of winter sunshine are shown (fig. 2).

The average January temperature in zone 1 is below 20° F.; in zone 2, 20° to 35°; in zone 3, 35° to 50°; and in zone 4, above 50°. Minimum temperatures seldom fall below freezing in zone 4, but they often fall below −30° in colder parts of zone 1. Summer temperatures are much more uniform throughout the Nation than are winter temperatures, the July average ranging from below 75° in zone 1 to above 80° in zone 4. Maximum temperatures are above 100° in all zones.

Sudden temperature changes may cause serious slumps in egg production. Exposure to hot sunshine or to high air temperatures and radiation from hot roof surfaces, especially when the relative humidity is high, may cause loss of hens from heat prostration. Laying houses should therefore be designed to provide both winter and summer weather protection.

Four Types of Laying Houses

Type 1. The wall-less, or wire-walled, house (figs. 3 and 4) is used in very warm climates. The front, rear, and sometimes the end walls are of wire netting. Lightweight, plastic-covered panels may be used to partly close the house in winter; or the walls exposed to winter winds may be covered with reinforced paper. Generally, there is

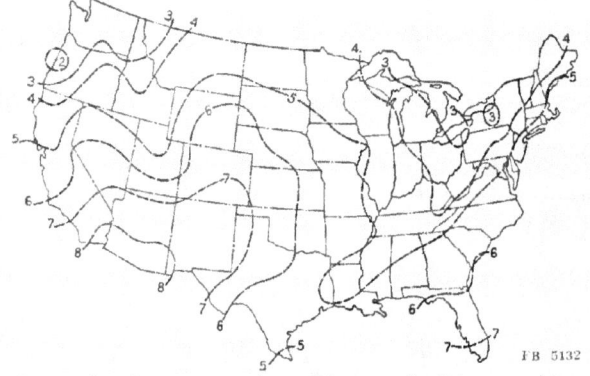

FIGURE 2.—Average number of hours of winter sunshine daily in December, January, and February.

FIGURE 3.—A wire-walled house in Arizona. The house faces south. Little sunshine enters in summer, but considerable enters in winter. The sunshade on the west may be lowered in cool, windy weather.

FIGURE 4.—A wire-walled house in North Carolina. Removable panels of plastic screens are used to close openings during cold weather, as shown. Note the continuous ridge ventilators and large end doors.

no wet-litter problem in these houses except during damp weather.

Type 2. The uninsulated building with large front-wall openings (fig. 5) is generally recommended by State agricultural colleges for flocks of all sizes in zones 3 and, to some extent, in zones 1 and 2. It is often provided with windows or curtains that can be closed during storms or cold nights. Under most conditions, open-front houses keep the litter dry and the eggs clean; but egg production may fall off seriously, and unprotected water fountains and eggs may freeze, in cold weather.

Type 3. The straw-loft, one-story house with ventilated attic, insulated walls, and some control of ventilation is recommended for zone 1 and for the colder parts of zone 2; with uninsulated walls, it is used in all parts of zone 2 and, to some extent, in zone 3. A typical straw-loft house is shown (fig. 6).

Type 4. The warmly built house with insulated walls and ceiling and controlled ventilation, about 24 feet square or larger in size, is recommended by most State agricultural colleges for zone 1. It is used for most large flocks in zones 1 and 2, and may be built either single-story or multi-story (fig. 7). Some newer houses of this type have large windows of insulating glass to trap more winter sunshine (fig. 8). Overhanging roofs are provided to keep out the sun in summer. Manufacturers of insulating glass can supply information on the overhanging roof required to suit the specific location.

Prefabricated poultry houses of various sizes are available complete with all poultry equipment.

Special Problems of the Small Flock

In northern parts of the United States, the small flock of less than 100 hens presents a special housing problem which is important because more than half of the flocks are of this small size. None of the four types of houses described, if fully exposed to the weather, is entirely satisfactory for so few hens, because the heat given off from their bodies is not enough to warm a house in cold weather if the house is adequately ventilated. On the other hand, flocks of less than 100 hens do not lay enough eggs to justify the expense of installing extra-heavy insulation or a heater in an exposed house, unless the eggs are worth far more than ordinary market prices.

One way to provide good conditions for the small flock in a cold climate is to keep them in a well-insulated room of a larger building. An alternative method, where a dry south slope is available, is to set the house into the hillside for protection. When this is done, floor and walls in contact with the bank should be waterproofed, and above-ground walls and roof must be insulated.

FIGURE 5.—An open-front laying house at the United States Agricultural Research Center, Beltsville, Md.

FIGURE 6.—Interior of a straw-loft house used in Missouri.

FIGURE 7.—A Connecticut type multistory laying house on a hillside location.

Special Problems of the Large Flock

In 1950, about one-fourth of all laying hens in the United States were in flocks of more than 400. Hens in large flocks are confined in the house at all times.

The present trend is toward houses 30 to 40 feet wide, though the width may be as much as 60 feet for large flocks. Wide houses are cheaper to build and are generally more comfortable for the hens in the winter than long, narrow houses of the same floor area; but if they are too wide, they may be hard to light and ventilate. Large drive-in doors built in the end walls are convenient at house-cleaning time, and they allow more air to circulate during summer. Wide houses are better suited than narrow ones for using litter stirrers, tractor scrapers, mechanized feeders, and other labor savers.

Multistory buildings (fig. 7) are warmer in winter and may be cheaper to build than large, single-story houses having the same floor area. They are popular in regions where level sites are hard to find. Their disadvantages include difficulty in moving feed, eggs, and litter to and from upper floors. A power-operated feed elevator or a bulk-feed truck may be used to deliver feed to the upper stories. A hillside location may permit the building of a ramp to the upper levels.

Space Requirements

The general trend is to limit the size of flock per pen to about 500 to 1,000 hens. General recommendations for laying houses without pits or slatted floors are shown in table 2. An additional half square foot per hen is desirable when hens are confined in hot weather.

Tiers of feeders and waterers may be placed above the droppings pit. A mechanical pit cleaner may be used to clean the pit frequently; this reduces the amount of evaporated moisture to be removed by the ventilating system. With this arrangement, some poultrymen allow only 3/4 to 1 1/4 square feet per hen.

Summer Comfort

In hot weather, free circulation of air through the house is essential for the comfort of the hens. Shading by trees, use of light-colored (or reflec-

FIGURE 8.—Interior view of a solar house, showing fixed panes of insulating glass. Note inlets above windows.

TABLE 2.—*Floor area for laying hens*[1]

Breed size	Hens in pen	Floor area per hen
	Number	*Square feet*
Small	25	4
	100	3½
	200	3
	400	2¾
	600–1,000	2½
Large	25	4½
	100	4
	200	3½
	400	3¼
	600–1,000	3

[1] In hot weather, allow more space per hen.

tive) roofing, and roof insulation aid in protecting the hens against extreme heat. In building zones 3 and 4, continuous ridge ventilators, floor-level windows, and large end doors are desirable (fig. 9).

Sizes of wall openings suggested for uninsulated, partly insulated, and well-insulated laying houses vary in the 4 zones. The openings indicated (table 3) are compromises between provision for free air movement in summer and protection against cold in winter. The sizes of these openings are calculated as percentages of the floor area. For example, in a 20- by 20-foot house, if an opening equivalent to 10 percent of the floor area is desired, the size of the opening would be 40 square feet.

To provide other aids to summer comfort:

Apply whitewash or white paint on galvanized steel; whitewash other dark-colored roofing.

Cover roof with straw, weeds, or other vegetation.

Double the watering space, and run water continually so that hens always have fresh, cool water.

When the outdoor-air temperature is above 90° F., place fans in house to increase air

FIGURE 9.—A 24- by 160-foot laying house with reflective roofing, continuous ridge ventilators, floor-level windows, and large end doors.

TABLE 3.—*Minimum area of wall openings*[1] *for summer ventilation, stated as a percentage of the floor area*

Type of house	Combined front and side walls				Rear walls			
	Zone 1	Zone 2	Zone 3	Zone 4	Zone 1	Zone 2	Zone 3	Zone 4
	Percent	*Percent*	*Percent*	*Percent*	*Percent*	*Percent*	*Percent*	*Percent*
Open front	(²)	10–15	12–20	18–30	(²)	1–4	6–10	8–20
Straw loft [3]	6–9	8–12	10–15	(²)	0–2	1–3	4–6	(²)
Well insulated [4]	4–6	6–10	(²)	(²)	0–1	1–2	(²)	(²)

[1] Not including area of entrance doors.
[2] Type not recommended for this zone.
[3] Walls not insulated.
[4] Includes straw-loft house with walls insulated.

movement by drawing air from the cool side of the house. Never circulate air from one pen to the next.

Use fog nozzles (fine-spray) inside the house to wet the hens. In humid areas this method should be used with caution. Avoid unnecessary moistening of the litter, as fermentation will heat the house.

Use evaporative coolers in low humidity areas.

Install sprinklers or porous hose on roof.

In an emergency, drive the birds out of house into woods or other cool shade.

In excessively hot weather, set blocks of ice in the house, and blow air against them with an electric fan.

To keep out summer sun from houses facing south, use adequate width of roof overhang; or sunshades as needed.

Winter Comfort

Poultry houses are ventilated in winter to remove excess moisture, odors, and airborne disease organisms, and to maintain dry litter for clean-egg production. The aim is to keep the relative humidity of the house air below 80 percent, and the litter moisture content below 40 percent. Temperature control in the house depends on construction, insulation, ventilation rate, and floor area per hen.

In building zones 3 and 4 (fig. 1) and in the warmer, sunnier parts of zone 2, uninsulated open-front construction is considered satisfactory by most producers. Drop curtains or shutters may be used to partly close the front for protection during exceptionally cold weather, but no special ventilation is needed for this type of house. The house may become quite damp when closed during cold weather, but it will dry out rapidly on warm, sunny days when the front is open.

In zone 1 and in the colder parts of zone 2, well-insulated or straw-loft houses are needed if the house temperature is to be kept above freezing during cold weather. Controlled ventilation is needed at that time to keep the house and litter dry.

Tests conducted by the Agricultural Research Service, Beltsville, Md. (1951-53), showed that an average of ½ to ⅔ pint of water per hen per day must be removed by ventilation, to avoid wet litter. Part of this moisture is given off in the hens' breath; the rest evaporates from droppings and from water spilled from drinking fountains.

The heat produced by the hens, plus that gained from sunshine and from fermentation of the litter, is not enough to maintain adequate warmth in the ordinary house on cold days, if the air is changed often enough to remove all the moisture produced. Consequently, during colder-than-average weather, the moisture content of the litter may build up to the "wet-litter" stage. The smaller the house, the harder it is to keep it dry in cold weather.

To reduce wetting of the litter:

Insulate the house well; this includes using storm windows or insulating-glass windows in the coldest climate, and keeping the glass clean to let in sunlight.

Use waterers that do not overflow and that are designed to prevent excessive water waste by the hens. A suitable wire-covered platform and drain, to dispose of waste water, will help. Minimum drain-tile size should be 6 inches, with a slope of 12 inches per 100 feet to an outlet. Some poultrymen connect a downspout from the roof to flush this drain.

Do not overcrowd the house.

Use a type of litter that does not pack readily and that hens like to scratch in. Keep litter depth at least 6 to 8 inches. Stir litter from time to time.

Remove wet litter around water fountain frequently, and replace with dry.

Where droppings boards are used or where droppings pits are equipped with mechanical cleaners, clean them frequently, thus removing much of the water in the droppings before it evaporates.

In houses having mechanical cleaners, place waterers and feeders over pit, and clean the pits often.

Take advantage of warm days and warm sunny hours to air the house.

When needed, supply additional heat to the house during cold or damp weather.

Systems of Ventilation

There are two types of ventilation in common use—the gravity, or natural, system and the fan-operated system.

The gravity system depends on cold outdoor air to force the warmer, and therefore lighter, air out of the house. This works well in the straw-loft house, with fresh air entering through windows opened 1 or 2 inches at the top, or through continuous narrow slots under the overhanging roof on the south side, or through special inlets. The exhaust air passes out through the straw, and through louvers in the gable ends of the house.

The gravity system is also used in insulated houses without straw-lofts, by providing ducts, or ventilator heads, for the escape of the exhaust air. In cold areas, flues must be extended well above the roof ridge, and they must be insulated to operate. If the inlets are well distributed and adjusted as the weather changes, this system can work well. However, considerable time may be required to adjust air inlets or windows almost every day during winter months.

In a modification of this system, one or more small electric motors, each controlled by a thermostat, are used to open and close the inlets or the exhaust ducts.

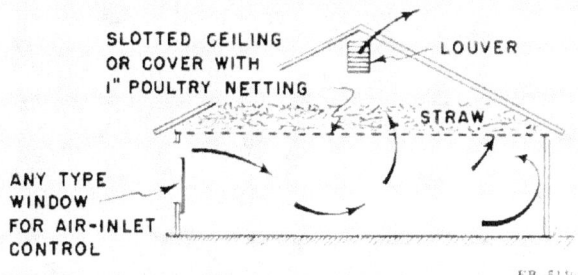

FIGURE 10.—Ventilation of a straw-loft house.

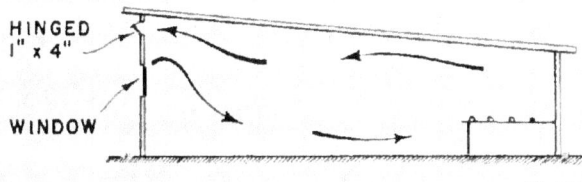

FIGURE 11.—Gravity ventilation of a fully insulated house.

Methods of obtaining draft-free gravity ventilation are shown (figs. 10 and 11).

In closed and well-insulated houses (not open-front or straw-loft), more positive control of ventilation may be obtained with a fan system. The capacity of the fan or fans should be adequate to supply at least 3 c. f. m. (cubic feet per minute) of air per hen at ⅛-inch static pressure. For a house with 250 hens, fan capacity should be 750 c. f. m.

To dry out the house after a cold spell, however, and to prevent too much temperature rise on warm days, 4 c. f. m. per hen is desirable. As this volume is much too high for cold days, there must be a damper to adjust the rate of airflow. In addition, it is necessary to do one of the following: Use a thermostat to shut off the fan when a marked temperature drop occurs (for example, a drop to below 35° F.); use 2 fans in a large pen—a small one to run continually, and a large one controlled by the thermostat (with the second arrangement, the small fan should have about one-third the capacity of the large one); or use a 2-speed fan having thermostat control, operated at low speed in cold weather and at high speed on warmer days.

Wall fans are cheaper and easier to install than ceiling fans. A wall type exhaust fan, set in a duct arranged to draw air from near the floor in cold weather and from near the ceiling on warm days, is shown (fig. 12). There is also a damper to regulate the airflow. When both a large and a small fan are used in the same room, the preferred location is side by side.

For a house up to 60 feet long, the fan may be set at the end of the house, away from prevailing winds. In a longer house, the fan or fans may be located at the middle of the rear wall of the house, immediately under the plate, and hooded against north winds. This arrangement draws fresh air through baffled openings on the warm side of the house, across its entire width, and exhausts the foul air on the cold side. In summer, the fan is reversed to draw cool, north-side air, and force heated air out the south side. There should, of course, be one fan or a pair of fans for each room that is enclosed by solid partitions.

For a gable-roofed house 40 or more feet in width, an exhaust fan may be located in the ceiling under the ridge of the roof to force the air out through a vertical duct.

A blow-in, or pressure, type of ventilation, which has been successfully used in Wisconsin and North Dakota in insulated houses with tight ceilings, is shown (fig. 13). Air drawn from the attic is warmed by sunshine on the roof and by heat leaking through the ceiling. The fan blades should be below the ceiling. The baffle under the fan is important to prevent drafts inside the house. Air leaves the house through narrow slots above or below the windows. With this arrangement, the fan should run continually or it should have self-closing louvers to prevent frosting caused by back drafting when it is not running.

It has been found that pressure (or blow-in) type ventilation prevents drafts around doors and windows of poultry houses and forcefully mixes cold air with warm air in the house.

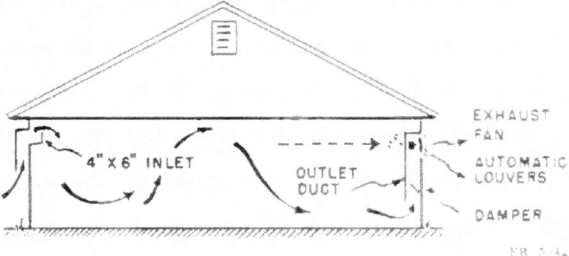

FIGURE 12.—A wall type exhaust fan-and-duct arrangement.

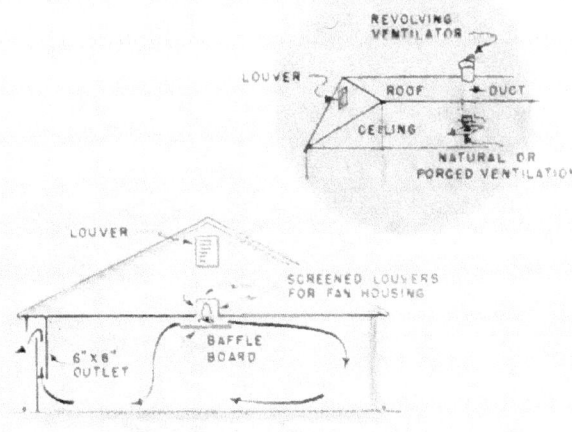

FIGURE 13.—Fan blowing air from attic into laying house. Inset shows alternative arrangement, with fan mounted in ventilation shaft.

Fan motors should be totally enclosed to keep out dust. They should also be equipped with thermal-overload circuit breakers, to prevent fire from starting in the motors, if the fan becomes frozen or stalled. Fans should be of the nonoverloading type, to prevent burning out of motors when louvers blow shut or fail to open. For efficient operation, the fan-motor housing and the blades should be kept free of dust and feathers. They should be screened with 1-inch poultry netting, to protect chickens and human hands from the propeller blades.

The inlets should give an even distribution of air throughout the house. A continuous slot, 1 inch wide by 12 feet long, or the equivalent, will supply sufficient inlet air for 100 hens. Devices to stop back-drafting are desirable.

An insulating straw loft is also an aid to ventilation (see next section).

Insulation and Wall Finish

Insulation is required to keep poultry-house temperatures above freezing in zones 1 and 2 and in parts of zone 3. Roof insulation helps protect the chickens from summer heat, but it should preferably be used in combination with light-colored or reflective roof surfaces.

Many types of commercial insulation are available, including fill, bat, blanket, board, and reflective. Any insulation that is fire-resistant and verminproofed may be used if it has the required insulation value.

Dry wood and various farm wastes, such as shavings, sawdust, ground corncobs, and chopped straw, may be used for insulation. Verminproofing and fireproofing treatments are available for these materials. These farm wastes have slightly less insulation value per inch of thickness than have commercial materials used for this purpose.

TABLE 4.—*Recommended minimum insulation values*[1] *for poultry houses*

Location	Size of house			
	20 by 20 feet		40 by 40 feet	
	Walls	Ceilings	Walls	Ceilings
Zone 1:				
Colder parts	8.0	15.0	5.0	10.0
Warmer parts	5.0	12.0	3.5	10.0
Zone 2	2.5	12.0	2.0	10.0
Zones 3 and 4		4.0		4.0

[1] Insulation value is defined as the number of degrees difference in temperature between the inside and outside surfaces of a wall, that will permit 1 British thermal unit (B. t. u.) of heat to pass through 1 square foot of the wall per hour.

The insulation values shown (table 4) are suggested as minimums for winter insulation of laying houses 20 by 20 feet and 40 by 40 feet in size. To maintain a given inside temperature, smaller houses need more insulation than larger ones. The insulation values provided by different types of construction are shown (fig. 14).

In insulated gable-roofed houses, it is more economical and satisfactory to insulate the ceiling instead of the roof, using fill or blanket insulation. The attic space must then be ventilated.

Several State agricultural colleges recommend the use of a straw loft, which serves both as insulation and as an aid to ventilation of the house. The usual depth recommended is about 10 to 12 inches of settled straw. Where space is available for more straw, 18 to 24 inches may be used. Any clean straw is satisfactory for this purpose, but chaffy straw or hay containing clover, alfalfa, or similar material is not suitable.

The space above the straw should be well-ventilated, usually by louvers in the gable ends, to prevent moisture from condensing under the roof. The area of the louvers at each end of the house should be $\frac{1}{2}$ to $\frac{3}{4}$ of 1 percent of the floor area. If the house is unusually long, roof or ridge ventilators may also be needed. No vapor barrier is used with a straw loft.

A suitable wall finish should be used inside the house. It should be able to withstand washing with hot water alone or with disinfectant in solution. The surface should be tough enough to withstand pecking wherever walls can be reached by the hens.

Vapor Barrier

A vapor barrier is needed on the warm side of an insulated wall, to prevent water vapor in poultry-house air from condensing in the insulation. Polyethylene film and asphalt-coated paper, 55-pound roll roofing, aluminum foil, or 2 coats of asphalt or aluminum-flake paint, applied on an unbroken surface, are satisfactory types of vapor barriers. Some commercial insulation materials have a vapor barrier attached.

Manufacturer's instructions should be followed in applying the barrier, and care should be taken to avoid tearing holes in it.

The outer-wall surface should be provided with small vents to allow the escape of any moisture that gets into the insulation. If vapor-resistant siding, such as roll roofing or asphalt siding, is used, ventilation to the stud spaces should be provided by means of small screened openings or slots (fig. 15, *A* and *B*).

The correct placement of the various materials in an insulated frame wall is shown (fig. 15, *C*). The siding is placed over 15-pound building paper or felt, which is not a vapor barrier but is used to prevent rainwater or snow from reaching the insulation. The space between the studs may be

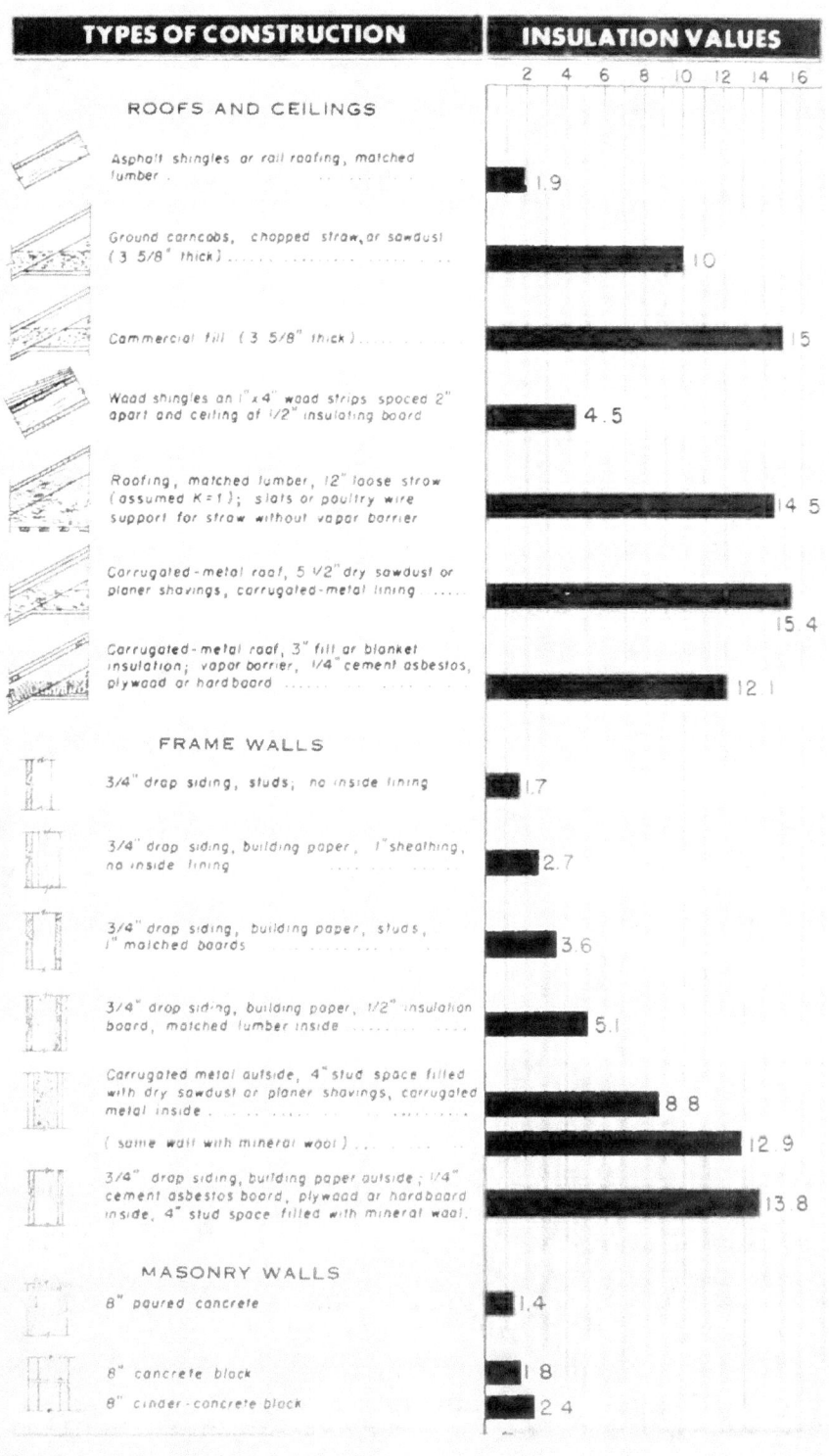

FIGURE 11.—Insulation values for common types of wall and ceiling construction. All attic spaces are considered ventilated.

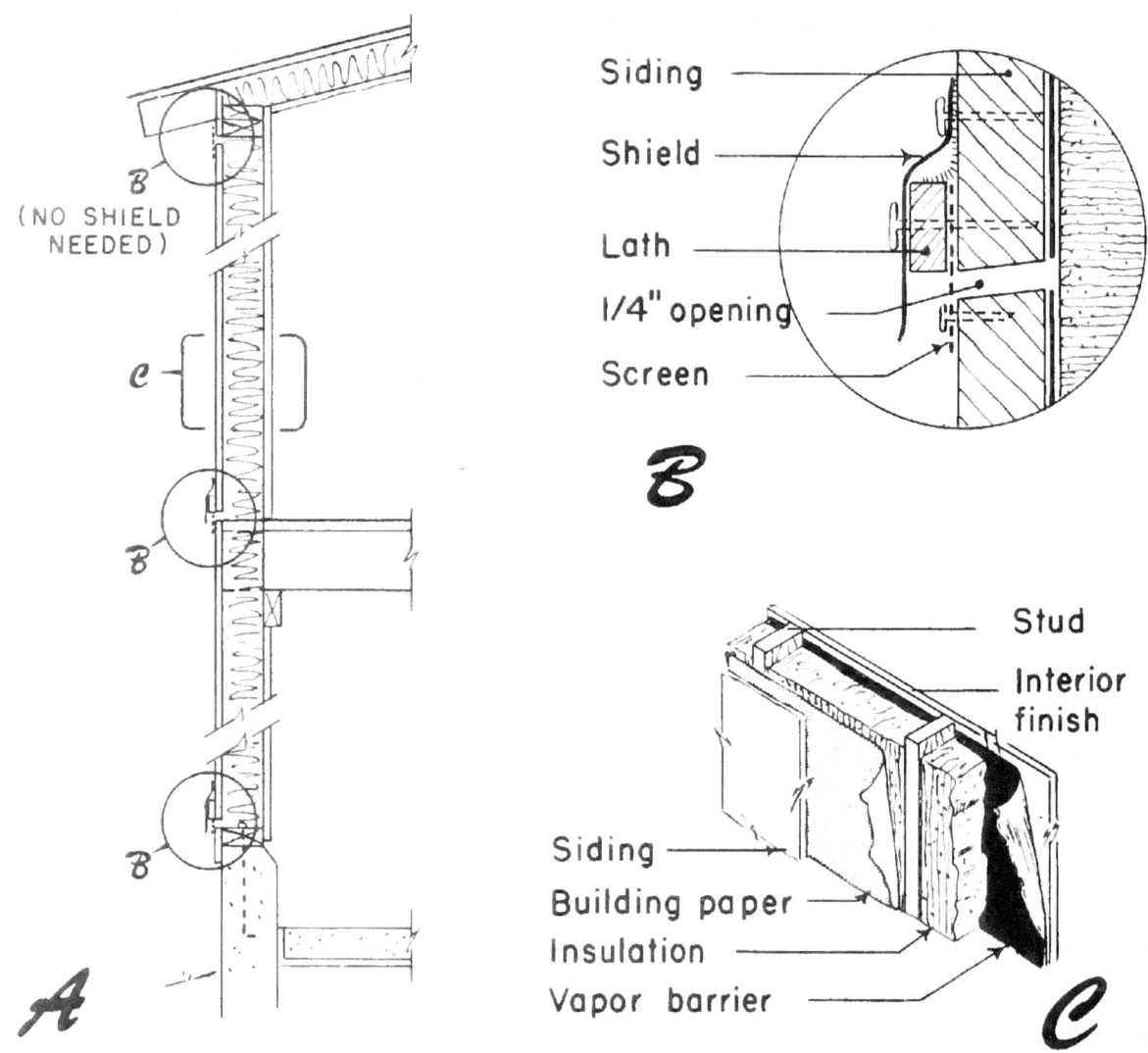

FIGURE 15.—Installation of insulation and vapor barrier: A, Ventilation to the stud spaces by means of small screened openings or slots; B, detail of small vent; C, correct placement of material in insulated frame wall.

insulated with farm wastes or commercial insulation. The vapor barrier is placed on the *inner surface* under the finish material, except when paint is used as the vapor barrier.

No vapor barrier is used with the straw loft (see preceding section). Such a barrier is used, however, for shed roofs that are insulated between rafters. In this construction, a vapor barrier should be placed below the insulation, and an airspace of at least 1½ inches should be left under the roof sheathing. This airspace should be ventilated by screened cracks under the roof boards at both ends of the rafters. The disadvantage of this arrangement is that roof leaks cannot be detected from the underside.

Building the Laying House

Construction Costs

Construction costs for materials and labor vary considerably in any given area. The cost of building a small house is considerably higher per hen than that of building a large house. In some localities, poultrymen reduce building costs by as much as 50 percent by using native lumber and other locally available materials and by employing family help.

The following estimates of cost in different regions of the United States are based on the prices of commercial building materials and poultry equipment and the wages paid hired labor for the construction work:

Section of the United States:	Construction cost per hen
Northeast (Connecticut)	$4.50–$7.00
Southeast (Georgia)	1.50– 3.00
Midwest (Missouri)	4.50– 8.00
Southwest	2.00– 6.00
Northwest	3.00– 6.00

Location and Orientation

For the farm flock, a site near other buildings saves time and travel and provides a certain amount of protection from thieves. Locating large laying houses 150 feet from other large structures reduces fire risk. The poultry house should be located where prevailing summer winds will not carry odors to the dwelling.

The site should be large enough to provide for expansion. Consideration should be given to location of service roads, water supply, electric lines, and future buildings and yards.

A site on relatively high ground, with a south or southeast slope and good natural drainage, is usually preferred. The foot of a slope where either soil or air drainage is poor, or where seepage occurs, is unsuitable. A location that is exposed to drainage from adjoining farms should be avoided because of the disease hazard.

If the house is on a hillside, the site should be graded to carry surface water away; and there should be a tile drain along the upper side at footing level, connected to an open outlet or to a dry well at a lower level. A house on a steep slope should not be faced uphill, as air drainage down the hill will make control of ventilation difficult.

For winter comfort and good litter management, the small or medium-sized house, in most climates, should face approximately south to obtain most sunlight and protection against northwest wind. Control of solar heat by shading the windows in summer is also more practical when the house faces south. However, an east front is preferred where prevailing winter storms are from the west. Wide houses are often placed with their long dimension north and south, to have equal lighting on both sides.

For disease prevention, older hens should be separated from immature ones by at least 200 feet. The farmstead should be planned so that chore routes, summer wind, and water drainage lead from young to adult stock.

Shade in summer is an advantage, but trees should be of the kind that shed leaves in the fall. A well-mowed sod greatly reduces radiation of heat from the ground into the building. **For fire protection, weeds and grass should be kept mowed around all buildings.**

In zones 1 and 2, windbreaks located 50 to 100 feet to windward of the poultry house help to keep it warm and to prevent snow from drifting around the house. Shrubbery around the house is not recommended, as it tends to trap heat in the summer and may shelter vermin.

Selecting a Plan

State agricultural colleges and the United States Department of Agriculture cooperate in preparing plans for poultry houses that meet requirements of the different parts of the United States. These plans may be ordered from the county agent, but *not* from the Department of Agriculture. Most county agents have catalogs that illustrate these plans.

Some manufacturers of building materials distribute poultry-house plans through lumberyards and other dealers. Other plans are available from publishers of poultry magazines.

The following points should be considered when selecting a laying-house plan:

Type of house. The house may be open, open-front, straw-loft, or insulated.

Size of house. (See table 1 for floor area needed per hen). Insurance companies recommend that the maximum capacity of the house not exceed 5,000 hens.

Width of house. The more nearly square the house, the less wall area there is in proportion to the floor area. Other things being equal, the wide or nearly square house is cheaper to build and warmer in winter than the long, narrow house. The maximum practical width is 50 to 60 feet, but many prefer houses not more than 40 feet wide.

Possible changes in farm enterprise. Headroom and spacing of columns should be planned so that the building may later be used to store machinery, to shelter other livestock, or for other purposes if a change in type of farming is made.

Spacing of partitions. Many poultrymen prefer that laying hens be penned in groups of not more than about 500 or 600. Sick and cull birds can be readily seen in flocks of this size.

Strongly built cross-partitions, to brace the building against high winds, should be spaced not more than 1½ times the width of the building. With this spacing, the area between brace partitions in a house 24 feet wide is 864 square feet. This space will accommodate about 250 large or 300 small hens. The area between brace partitions in a 50-foot-wide house would be 3,750 square feet, which is enough for 1,500 small hens. One or more wire partitions may be used to divide this space.

Wall openings needed for summer ventilation. (See table 3).

Insulation required. This is determined by the type and size of house desired, and the climate in which it is to be built.

Size of doorways, spacing of supports, and height of ceiling. The dimensions and spacings of these items depend on whether or not tractor equipment is to be used for cleaning the house. Minimum width of door for a tractor is about 8 feet and the height about 7 to 8 feet, depending on the design of the tractor.

Selecting Building Materials

Wood construction on concrete or masonry foundations has been used satisfactorily for poultry houses for many years. The same is true of concrete and cinder blocks and clay tile. Use of metal roofing and siding has increased since World War II. More recently, the use of pressure-treated poles and planks for foundation and wall framing is becoming popular. In zones 3 and 4, the wall sometimes consists only of wire netting stretched on the framing. In winter this may be covered with reinforced kraft paper or with a durable translucent plastic.

Any one of these types of construction is satisfactory if it provides: Structural strength to resist wind and snow loads and to carry the weight of hens, feed, litter, and other contents; insulating value sufficient to maintain the desired temperature in the house; and reasonable initial cost and upkeep expense.

Prefabricated houses of several types and sizes are on the market. A prefabricated house may meet requirements more economically than one built on the site, and it can be erected much more quickly. Since it is not economical to make changes in a house of this type, the plan and description should be studied carefully before purchase.

Building Details

If a suitable plan for the house has been selected, the blueprints should show the necessary construction details. If the design is changed, a qualified engineer or builder should be consulted to see that the construction is strong enough. For example, if a thin concrete slab is substituted for a wood floor, or if a feedbin is installed on an upper floor, the joist, column, and footing sizes should probably be increased. The points discussed in the following paragraphs are emphasized because they are often neglected.

Severe wind storms—for example, "Hurricane Hazel" in 1954—have destroyed many poultry houses on farms where dwellings and barns were left unharmed. Examination of many poultry-house failures showed that weak construction of foundation and framing and lack of cross-partitions in long houses were largely responsible for these failures. Details showing good construction practices for hurricane areas may be found in U. S. Department of Agriculture Information Bulletin 144, "Preventing Storm Wind Damage to Farm Buildings."

Footing and foundations

Foundations must have sufficient weight or grip on the ground to hold the building down; 18-inch depth should be the minimum in areas subject to windstorms. Greater depth may be needed, depending on local soil and frost conditions. For a permanent structure, a continuous footing (fig. 16) of poured concrete is desired to spread the weight of the building evenly. Steel reinforcement may be necessary where firm soil and good drainage are not available.

In frame houses, anchor bolts should extend into the footing if the foundation wall is of concrete blocks. If the wall is of poured concrete, ½-inch bolts, set at least 6 inches into the top of the concrete, and spaced 6 feet, are generally satisfactory, except in hurricane areas. Large washers should be used on top of sills.

Piers for posts and columns should be carefully made of concrete, because a relatively small bearing area carries several times the load of an equal area along the foundation wall. The footing sizes and depths shown on plans approved by the State agricultural college should be followed.

The size of footings for interior posts is extremely important, both in supporting the weight of the building and its contents and in resisting uplift during windstorms. The top of the pier should be at least 3 inches above the litter and usually 8 to 12 inches above the floor. Dowels (pegs) should not be depended on to hold posts

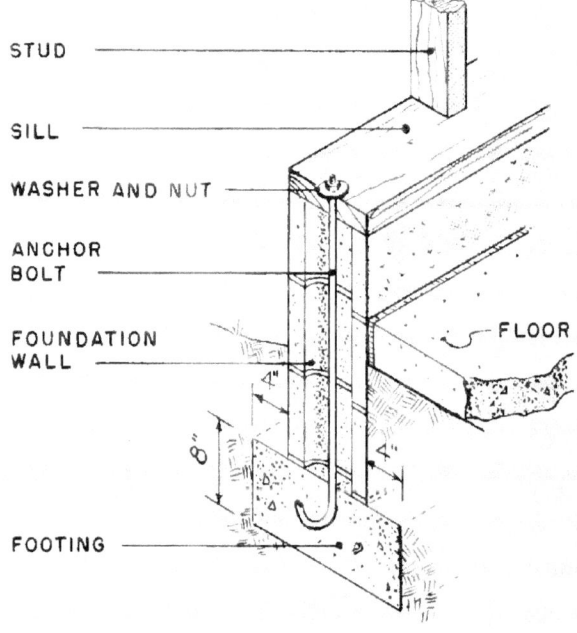

FIGURE 16.—Detail of foundation and footing. Note that anchor bolts extend into footing.

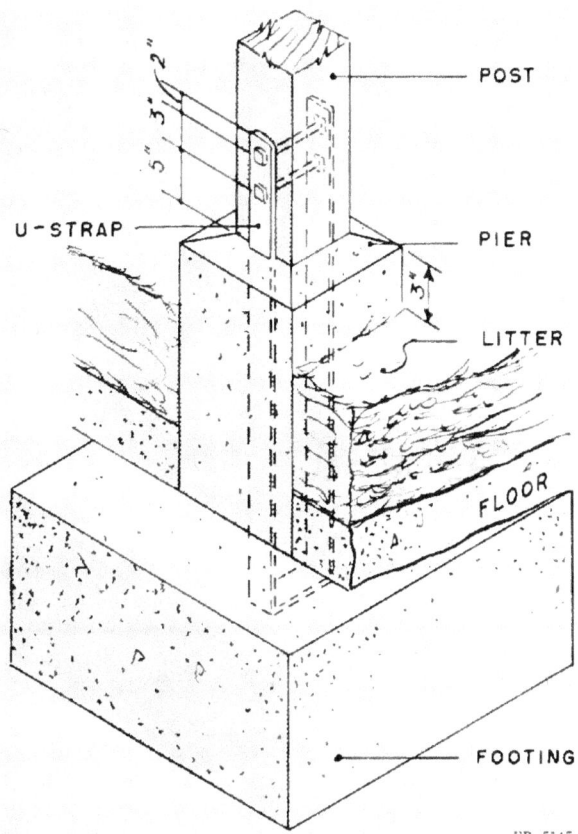

FIGURE 17.—Pier and anchor for posts. Note U-strap extending into footing.

on piers. A better method is to imbed a steel U-strap into the footing and lag-screw or bolt it to the post, as shown (fig. 17).

Floors

A floor in contact with the ground is warmer in winter and cooler in summer than one that is laid above a shallow "crawl space" so that its underside is exposed to outdoor air. Concrete is generally preferred, as it is smooth enough to permit easy removal of litter, can be disinfected, and helps keep out rodents.

Dirt, gravel, and limestone floors are widely used but cannot be satisfactorily disinfected. Since they are cheap, they enable an inexperienced poultryman to keep his investment as low as possible until he can afford to lay concrete.

Floors of concrete, dirt, or similar material should be raised 8 or more inches above the surrounding ground. Where seepage is a problem, some State agricultural colleges recommend placing a 4- to 6-inch gravel fill under the floor for drainage. A layer of 55-pound roll roofing or polyethylene film, overlapped about 2 inches at the joints, makes an effective moisture barrier between the floor and the subgrade. In building zone 1 and in the cooler parts of zone 2, a 1-inch-thick strip of waterproof insulation between the floor and the outside wall helps keep the house warm and reduce condensation on the floor (fig. 18).

A 3- to 4-inch thickness of well-mixed concrete (1 part cement, 2 parts sand, and 4 parts gravel), laid on a well-tamped base, is usually sufficient to carry the tractor and manure spreader used in cleaning the house. If the floor is to be scrubbed, it should be sloped toward one or more doorways for drainage.

A wooden floor is generally used for the upper stories of multiple-story houses. Flooring should be used which does not splinter badly when scraped, and which resists rot. Edge-grain fir and yellow pine are suitable woods for this purpose. Sometimes a 2-inch thickness of concrete is laid over wood boards or corrugated sheet iron. This adds about 25 pounds per square foot to the load on the floor joists, posts, and footings; the construction should therefore be strong enough to carry this additional load.

Wood framing

Long houses should be strengthened against wind by solid partitions spaced not more than 1½ times the width of the building. End walls and solid partitions must be braced either by diagonal bracing (fig. 19) or by boards applied diagonally. In multistory buildings, 1- by 4-inch braces should be used in the top story, and 1- by 8-inch braces for lower stories.

Connections between the foundation and the walls must be sound. Anchor bolts should be used through sills; studs should either be toenailed to sill with five 10d nails, or tied with 22-gage steel strapping or special fasteners.

The connections between wall and roof are important. Either steel strapping or commercial steel anchors (fig. 20) should be used.

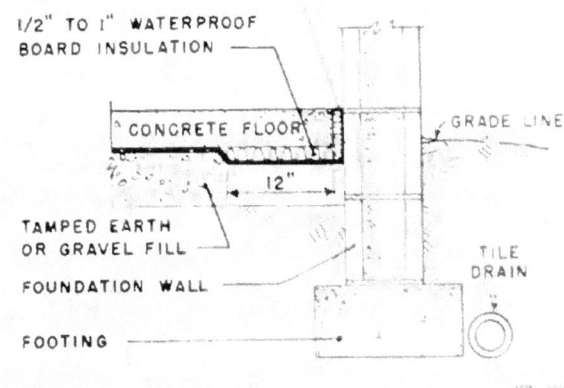

FIGURE 18.—Detail of floor construction, showing moisture barrier of 55-pound roll roofing and edge insulation, for zones 1 and 2.

Roof covering

The roof is exposed to a greater sunshine-heating load than any other part of a building, and it also provides the best heat-radiating surface at night. On a still, clear day, the temperature of a roof with a black upper surface may be 50° F. or more above the outdoor-air temperature. Where summer comfort of the hens is most important, the surface of the roofing material should be white or light-colored; or a material that reflects heat, such as aluminum, should be used. If a dark-colored roof has already been installed, it may be whitewashed before hot weather begins.

For roofs with little slope. Built-up bituminous roofing can be used on roofs that have very little slope. It is often applied by roofing contractors. Asphalt-roll roofing with cemented joints is usually used on roofs that have from 2- to 4-inch slope per foot. It may also be used on roofs having steeper pitch.

Single-ply roll roofing is most commonly used for nearly flat roofs, but the life of such roofs is usually short. Wide-selvage roll roofing that gives double coverage, with the upper ply cemented to the lower, is considerably more durable than ordinary roll roofing. However, it costs about twice as much.

When built-up roofing is used on an insulated house, the space between the roof and the insulation should be ventilated, to allow escape of moisture that leaks through a punctured vapor barrier, or through other openings or cracks in the building.

For roofs with steep slope. For roofs that slope 4 inches or more per foot, asphalt shingles and sheet-metal roofing can be used. Asphalt shingles are more pleasing in appearance than the roll roofings. The lock-down type of shingle is more

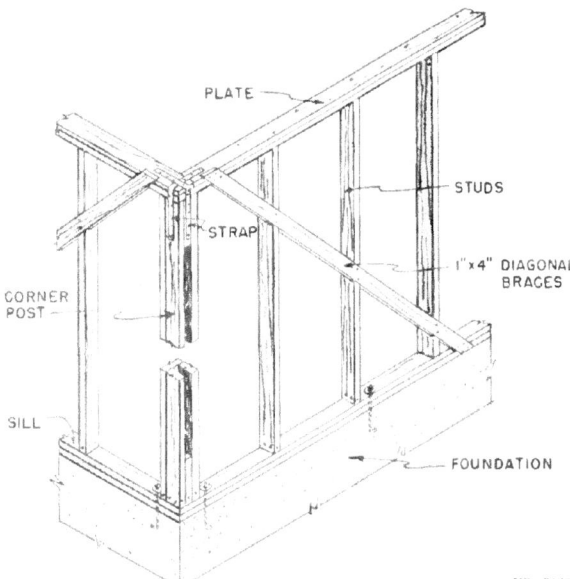

FIGURE 19.—Diagonal bracing and steel strapping used to strengthen end walls. Similar bracing should be used in solid partitions.

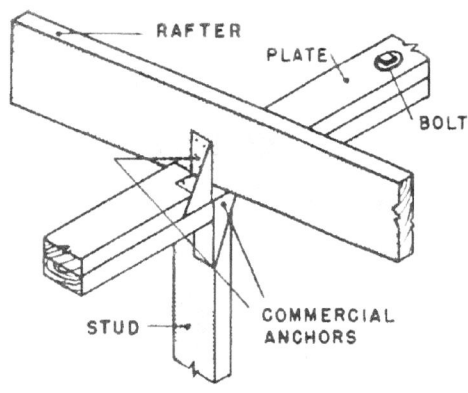

FIGURE 20.—Use of commercial anchors to tie plates to studs and rafters to plates. The upper half of the plate should be bolted to the lower part with ½-inch bolts, placed 4 feet apart; or steel strapping may be used.

Connections of center post to girder and girder to rafters, using steel strapping, are shown (fig. 21). Failure from wind damage often occurs here, especially in buildings with roofs that are nearly flat or shed type.

As masonry-block walls are often damaged by storm winds, special care should be given to quality of workmanship and to tying the walls to floors and roof. U. S. Department of Agriculture Information Bulletin 144, "Preventing Storm Wind Damage to Farm Buildings," shows simple means of strengthening both masonry and wood-frame buildings.

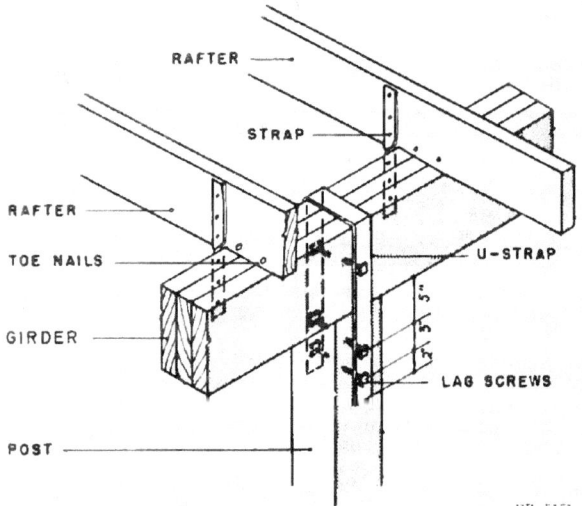

FIGURE 21.—U-strap, 2¼- by $^3/_{16}$-inch, tying girder to post. Fasten strap to post with ⅜- by 2½-inch lag screws, as shown. Use plumber's strapping or commercial anchors to tie rafters to girders.

resistant to wind damage than the rectangular-strip shingles. The tabs of rectangular shingles should be cemented down.

Aluminum-roofing sheets not less than 0.024 inch thick, or galvanized-steel roofing sheets of 26-gage material (preferably painted white on top) make a good roof, if nailed down securely. If 0.019-inch aluminum or 28-gage steel is used, such material should be laid over tight sheathing.

Manufacturers' recommendations should be carefully followed for best results in applying roofing materials.

Windows

There are numerous satisfactory designs for windows. None are perfect, and all require maintenance.

One type often used for the laying house is the counterbalanced window, outside a masonry wall (fig. 22).

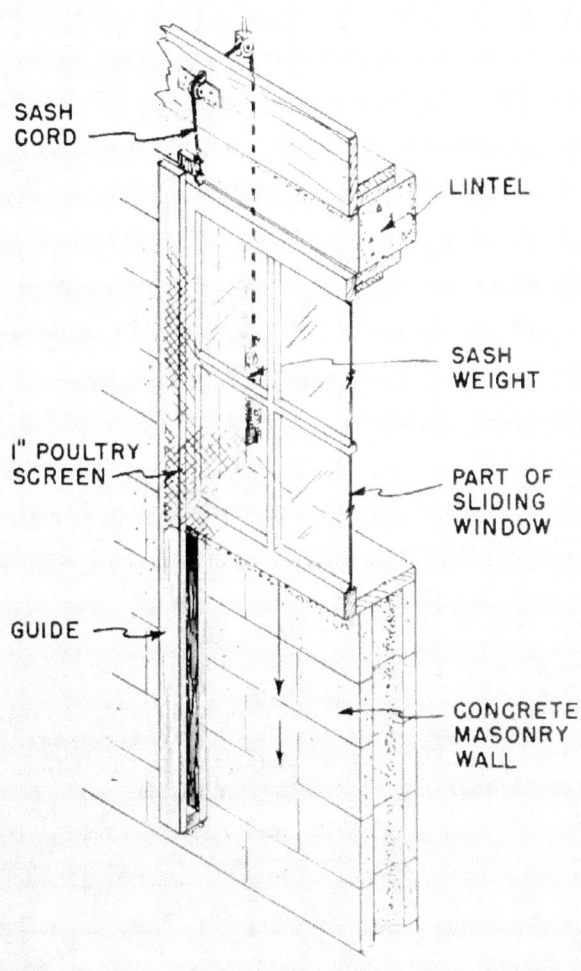

FIGURE 22.—Counterbalanced window hung outside of masonry wall.

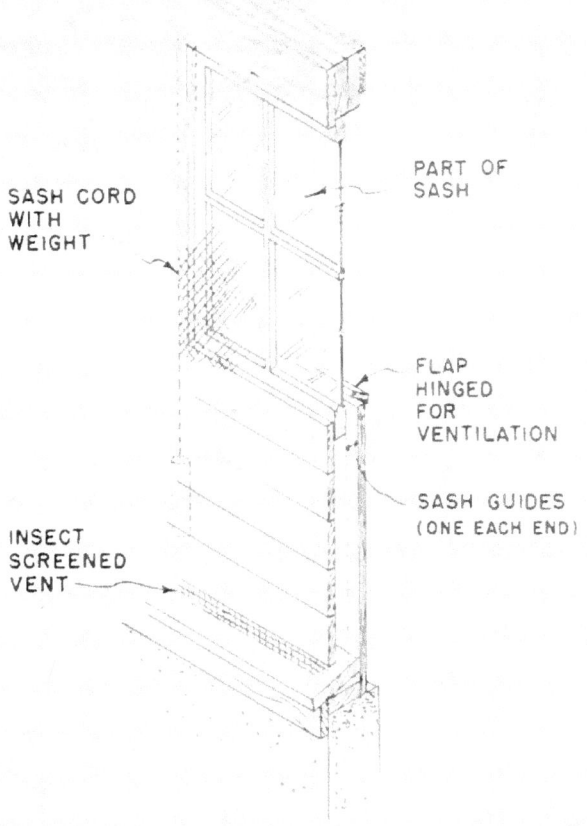

FIGURE 23.—Counterbalanced window sliding into frame wall.

Windows hung on the outside require more regular maintenance than those sliding into the wall (fig. 23). Some poultrymen are now using aluminum windows. It has been observed that these require less maintenance than windows with wood sashes.

In the warmer climates, muslin of heavy unbleached grade or a glass substitute is sometimes used in place of glass on a wooden frame. Cloth or fabric underneath the molding strips should be waterproofed to prevent rotting. The bottom should be tacked to the frame but not covered with molding.

Windows on the north side of the house in zones 1 and 2 should be weather-stripped to stop drafts. Windows shown in figures 22 and 23 cannot easily be weatherstripped. Storm windows or windows of double glass are desirable in zone 1. If windows are not needed for light, wood or plywood shutters may be used.

Window openings should be screened with 1-inch-mesh poultry netting to keep out wild birds. Screening on the inside prevents hens from roosting on the sills.

15

FIGURE 24.—Shutters hinged to serve as shades.

Shutters hinged at the top and opening outward help to keep out summer sunshine. Such shades (fig. 24) can be made of metal, pressed wood, plywood, or other suitable material.

Interior Details

For large flocks, the time required to feed and water, gather eggs, and observe the hens may vary from 20 minutes to 2 hours or more per day per 1,000 layers. Persons tending a flock may walk from 100 to 500 miles per year. As much as 50 percent of chore time may be spent in handling and collecting eggs. These figures do not include large cleaning jobs that must be done occasionally. It is important, therefore, to consider carefully the selection and arrangement of equipment, labor-saving devices, and other interior details.

A schematic diagram of an expandible layout for saving labor and time in a large laying house is shown (fig. 25). The nests are located near the egg-handling and egg-holding rooms.

Nests

Most commercial type nests are easily cleaned, and ease of cleaning is important. Two types of commercial nests are in general use—individual and community.

Individual nests are just large enough for one hen (fig. 26). Some have an egg roll-out floor and tray which may also be incorporated with a belt egg conveyor.

The usual dimensions of an individual nest are 10 to 12 inches wide, 12 to 14 inches high, and about 12 inches deep. From 1 to 2 inches of

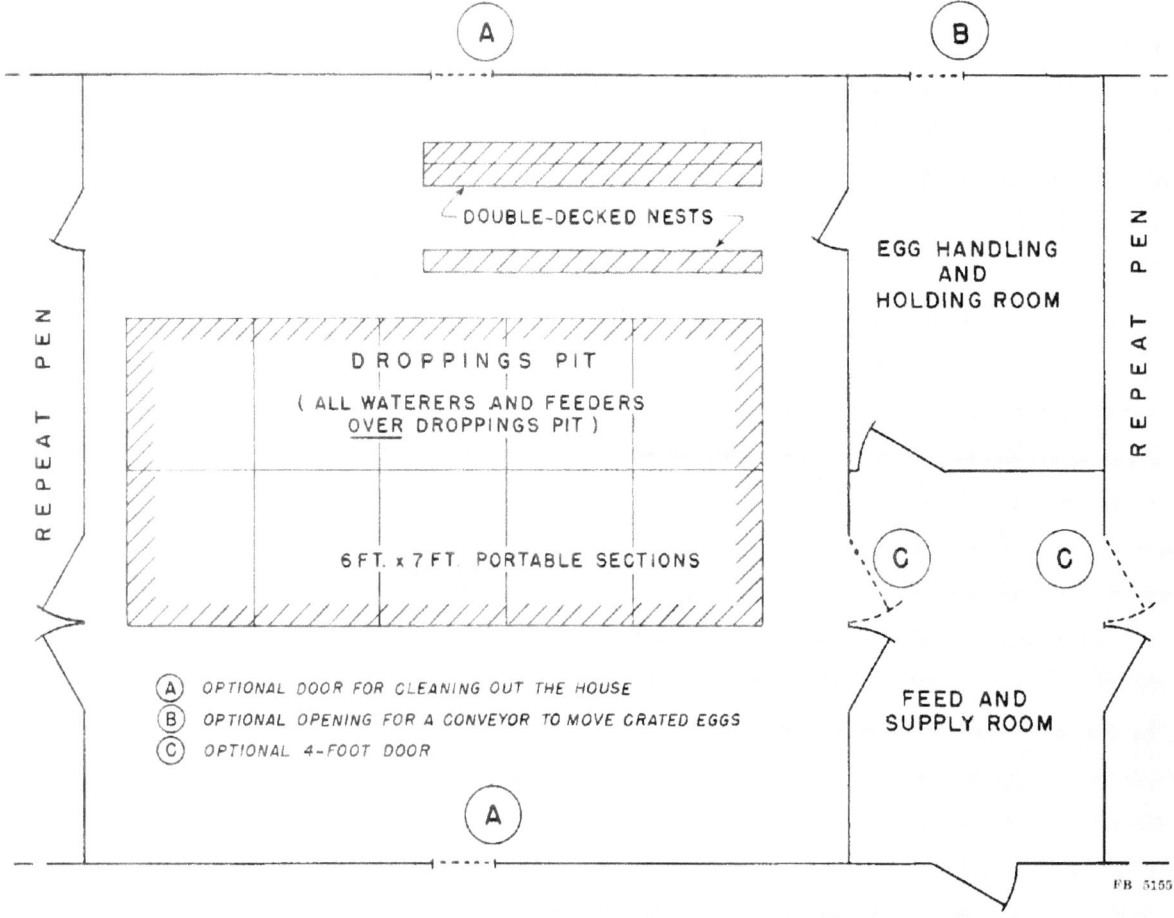

FIGURE 25.—Schematic diagram of an expandible layout for saving labor and time in a large laying house. Portable droppings-pit section may be 1- x 2-inch wire cloth or wood slats.

FIGURE 26.—Prefabricated individual nests against wall; trough type feeders in center of room; grit and shell boxes attached to post.

FIGURE 27.—Prefabricated metal community nest, with sloped floor to roll eggs away from hens.

FIGURE 28.—Homemade community nest: A, Front entrance; B, rear view, with eggs rolled away from hens.

nesting material is needed for each nest. One nest is needed per 4 or 5 hens. A 4- to 6-inch landing, or jump, perch in front of the nest, covered with either ½-inch-mesh hardware cloth or 1- by 2-inch welded wire, will help keep the nests clean. Some poultrymen keep hens from roosting in the nest at night by hinging the perch to fold against the opening of the nests.

Prefabricated community (or colony) nests, which accommodate several layers at one time, are shown (figs. 27 and 28). These nests are available with roll-away floor and egg tray (fig. 27). Also, belt egg conveyors can be had.

Satisfactory roll-away floors for community nests can be made by the poultryman by using ½- by 1-inch, 16-gage, welded-wire fabric having a 1¼ to 1½-inch slope. A nest 2 feet wide by 4 feet long is sufficient for about 40 hens. One entrance, 8 inches square, is needed for each 2- by 4-foot nest.

As several hens use the community nest at one time, ventilation is important. There should be at least a 1-inch crack, or equivalent, at the top for escape of heated air.

The trap nest is not ordinarily used by market-egg producers. Trap nests enable poultrymen to keep individual egg-laying records of the hens. These nests require frequent egg collecting to free the hens. One nest for about 4 hens is sufficient.

Feeders for mash, oystershell, and grit

When the hens are hand-fed, at least 40 linear feet of trough space is required per 100 hens. A strip of 1- by 2-inch-mesh welded wire may be cut to lay on top of the feed, to prevent the chickens from spilling it from the trough.

Readymade metal feeders are generally used. A trough type of metal feeder, suitable for mash or pellets, is shown (fig. 26).

To produce eggs with good shells, the hens must be provided with limestone or oystershell. Normally, 100 hens consume about 1½ to 2 or more pounds of oystershell per day, depending on air temperature, feed consumption, and egg production. The consumption of calcium increases during warm weather, and an ample supply should be kept in the hopper at all times. A 12-inch hopper for grit and another of the same size for oystershell are sufficient for 100 hens. Grit and shell boxes attached to post are shown (fig. 26).

Mechanical feeders (fig. 29) have become popular in the past few years. These are helpful to poultrymen in handling large flocks. There is some disagreement as to the best size of flock for economical use of such feeders. Improvement of feed-hopper design has greatly improved the efficiency of these machines. Local feed dealers may contact mechanical-feeder manufacturers for assistance in laying out the equipment.

FIGURE 29.—Automatic water fountains and mechanical feeders mounted over a utility pit

Waterers, and piping of water to fountain

Buckets, pans, and other hand-filled waterers are used for small flocks. When the flock is large, much time is saved by having the water piped to fountains. Watering devices may be automatically controlled, or they may be the continuous-flow type.

The amount of water required depends on the type of waterer. The float type system takes 4 to 5 gallons per 100 hens; in summer it requires up to 6 or 9 gallons. The constant-drip system and the continuous-flow system take considerably more.

Automatic-control type waterers include fountains that maintain a specified depth or weight of water. A tank is often connected to the pipeline to furnish a reserve supply if power fails and to provide a convenient means of adding medication. A round metal fountain is shown (fig. 29) and a trough type fountain (fig. 30).

In the continuous-flow type, the water level is kept just high enough in a V-shaped trough to permit the hens to drink. Fluctuating pressure in the pipeline makes it difficult to adjust the needle valves in a continuous-flow system. To overcome this difficulty, the waterers may be fed from a gravity tank filled from a pressure line with float-valve control. If the rate of flow is fast enough, the water does not freeze. An adequate drainage system is needed to handle the continuous flow of water.

For 100 hens, a 5-foot trough accessible on both sides is required with either system. In summer weather and in warm areas, watering space may have to be doubled.

A fountain should be placed either on a platform made of 1- by 2-inch-mesh welded-wire floor, 30 to 36 inches wide and raised 3 to 4 inches higher than the expected litter depth, or over a drain (fig. 30). Watering with this arrangement will lessen the amount of wet litter surrounding the fountain.

Except in warm locations, or in houses that are so warmly built that there is no danger of freezing inside, waterers need protection from frost. Thermostatically controlled heaters are available for maintaining ice-free fountains.

Water pipes should be laid underground for frost protection, to keep them out of the way and to keep water cool in summer. The pipes may be made of galvanized steel or plastic, or rubber hose may be used. Copper pipes are corroded by ammonia fumes and therefore should not be exposed to air in the house. Galvanized-steel pipes will corrode when laid in cinder fill.

Either a ½-inch galvanized steel pipe or a ⅜-inch plastic pipe is large enough to supply water for a large number of hens. If water is used for washing down the inside of a house, for roof sprinklers, or for special nozzles for fogging, a pipe ¾-inch or larger will be needed, depending on the size of the house.

FIGURE 30.—Automatic trough-type fountain mounted over a drain covered by a wire platform. Note electric heating cable wound around the water pipe.

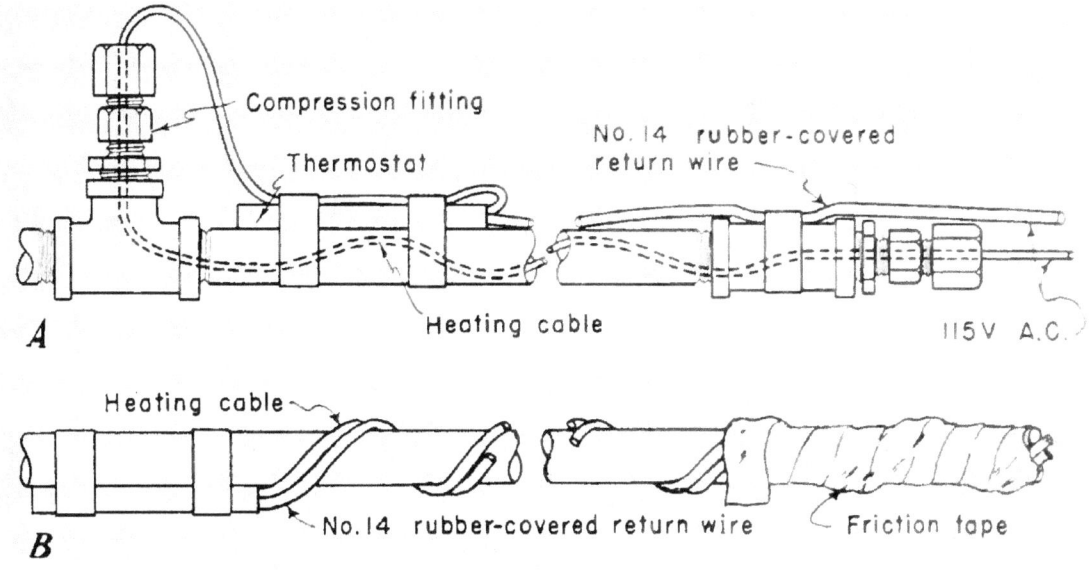

Figure 31.—Two ways to use heating cable with water pipes: A, Heating cable inside the pipe and rubber-covered return wire outside the pipe; B, heating cable and return wire outside the pipe, protected by friction tape or insulation tape.

Water pipes exposed to low temperature may be protected from freezing by using an electric heating cable. Electric heating wire installed in the waterline is shown (fig. 31, A). This heating wire is about ⅛ inch in outside diameter and is covered with water-resistant material. Plastic-covered soil-heating wire is satisfactory. Two special compression fittings are needed at each end of the heating wire, and a thermostat is needed to operate the heater. The wire shown on top of the pipe is an ordinary No. 14 rubber-covered-copper return wire. The pipe must be grounded as protection against possible electric shock.

An alternative method, in which about twice as much electric energy is used as in the one just described, is shown (fig. 31, B). In this method, more heating wire is also required. The heating and the return wires should be spiraled in parallel, as shown. The wires are covered with friction tape or pipe insulation.

The amounts of electrical energy needed to maintain ice-free water in pipes at various house temperatures are given (table 5). These are the values for the 2 methods illustrated (fig. 31, A and B). For example, if 220 feet of ½-inch pipe in a house needs frost protection at an expected low house temperature of 20° F. and the heating wire is installed inside the pipe, a total of 185 watts $\left(\frac{220}{100} \times 84\right)$ will be needed. If refrigeration insulation tape is used to wrap the pipe, less energy will be used. The agricultural engineer of the local power company should be consulted about the length, resistance, and wire size of heating cable.

Roosts, droppings pit, droppings board, and utility pit

The droppings pit, with roosts above it, is designed to accumulate droppings for several months. A typical design is shown (fig. 32). If the pit is boxed in, the maximum length of section should be 7 to 8 feet, for easy culling and cleaning. Some poultrymen prefer to keep pits, when located against the wall, less than 6 feet wide for easy culling. When pits are located in the center of the house, the width should be not more than 14 feet. For easy handling of the pit enclosure, the roosts and walls may be constructed separately so that they can be moved.

Table 5.—*Amounts of electrical energy needed to maintain ice-free water in pipes at various temperatures (University of California)*

Method of heating	Energy per foot per °F below 32°	Energy per 100 feet of pipe at various house temperatures			
		0° F	10° F	20° F	30° F
	Inch	*Watt*	*Watts*	*Watts*	*Watts*
Cable through pipe					
½	0.07	224	154	84	14
¾	.10	320	220	120	20
Cable spiraled around pipe					
½	.17	544	374	204	34
¾	.20	640	440	240	40

A typical design of droppings board is shown (fig. 33). The open space under the board provides additional floorspace for the hens. The roosts are generally hinged to the wall and raised when cleaning the droppings board.

Roosts (perches) should be of 2-inch stock, rounded or beveled on the upper edges. For use in hot weather, perches should provide 8 inches of space per hen for small breeds and 8 to 10 inches for large breeds. Perches are usually spaced 13 to 15 inches apart.

Some poultrymen use multiple-tier roosts over droppings pits to increase the number of layers housed (fig. 34). Feeders and waterers are located over these utility pits. Over 75 percent of the droppings can be collected in the pits, and frequent removal of droppings will greatly decrease the amount of moisture that accumulates in litter. A complete prefabricated unit of feeders, waterers, nests, and mechanical pit cleaners are now available.

Feed room

The feed room may include space for supplies and feedstuffs, workbench, egg handling, and egg holding. The egg-handling and egg-holding space should be walled off to keep dust from settling on the equipment, packing cases, and supplies, and to aid temperature control.

To calculate requirements and space for feedstuff, it can be assumed that in 1 day 100 hens will eat 30 pounds of all-mash or grain-and-mash diet. If the diet is half grain and half mash, space for 15 pounds of each is needed per 100 hens. Thus, for 10 days, 300 pounds of all-mash, or 150 pounds each of grain and mash, is required. To offset feed-delivery difficulties during bad weather, space for at least an extra 2 days' supply is desirable. To calculate storage space, use table 6.

In many areas, feed is delivered in bulk to save time, cost, and handling of bagged feed. Roads leading to the laying house should be surfaced to withstand 20 to 22 tons of feed delivered in trucks. A road clearance of 12 or 13 feet and space in which to turn around are required. The feed dealer should be consulted regarding location and height of bins and other details of bulk-feed delivery. Poultrymen in grain-producing areas often find it convenient to store an entire year's supply of feed at harvesttime.

Manufacturers' plans of feedbins constructed of plywood, pressboard, or other materials are available through lumber dealers. In many areas, prefabricated metal bins may be purchased. A few

FIGURE 32.—Droppings pit built along rear wall in a prefabricated metal house.

FIGURE 33.—Droppings board built along rear wall.

manufacturers make steel hopper bottoms for which farmers can build walls, roofs, and supports. Some agricultural county agents have plans for bulk bins. For unusually large bins, however, an engineer should be called upon to design the structure.

Bins with 1 vertical side (3-way-sloped bins) give less trouble from bridging than bins with 2 or more vertical sides. A 4-way-sloped bin is shown (fig. 35). Mash feeds require a more steeply sloped hopper bottom than grain. An antibridging device is needed to permit the mash to flow out of the bin.

The following construction details for bins are especially important:

> Exterior walls and roof should be tight, to prevent rain or snow from entering and to permit effective fumigation if needed.
>
> If bin is exposed to sun, white or aluminum-colored paint or material should be used.
>
> Smooth inside walls are needed.
>
> Bins should not be located on an upper floor without proper support underneath.
>
> Spouts for emptying bins used for storing mash should measure about 10 by 12 inches. A metal door of flat steel, about ⅛ inch thick and sliding in a metal groove, is most satisfactory for large bins.

FIGURE 34.—Multiple-tier roosts, feeders, and waterers over droppings pit.

FIGURE 35.—A 4-way-sloped bin in California. Elevator shown at top transfers feed to hopper bin for easy filling of feed carts.

TABLE 6.—*Weight and space requirements for various feedstuffs* [1]

Material	Weight per cubic foot	Space per ton
	Pounds	Cubic feet
Grain:		
Barley	40	50
Corn, shelled [2]	44	45
Grain sorghum or milo	41	49
Oats	28	72
Soybeans	46	44
Wheat	48	42
Mash:		
Finely ground	29	69
Coarsely ground	34	59
Crumbled	34	59
Pelletized, hen size	37	54
Middlings, loose	25	80

[1] 1 bushel = 2,150 cubic inches, or 1.24 cubic feet.
[2] Ear corn occupies about twice as much space as shelled corn.

Egg-handling room or workroom

On most small farms, a separate workroom for cleaning, grading, packing, and holding eggs may be conveniently located in a cool, clean home basement. If more than 10 cases of eggs are produced per week, the workroom is best located near or in the poultry house. The room should be so located that it can easily be expanded later at small cost. One end of the feed room is a convenient place if space is available, but the egg-handling room should be walled off to keep out dust.

A smooth floor is desirable for ease of cleaning and moving wheeled equipment. For concrete floors, a 1/8-inch-per-foot slope to a drain is satisfactory. Floors should be kept as dry and clean as possible, as wet and dirty floors are unsanitary.

If additional humidity is needed in the room, inexpensive electric humidifiers are available.

The following approximate floor areas are suggested for the workroom, to permit convenient arrangement of equipment for egg grading and packing, for moving filled cases of eggs, and for storing empty egg baskets and a few days' supply of empty cases:

Laying hens (number)	Workroom area (square feet)
1,000	80–100
5,000	100–150
10,000	200–225

Since floorspace requirements for automatic egg cleaners and graders vary, a simple scaled cutout of the particular make of equipment will help in planning the layout. A typical layout is shown (fig. 36). The placement of egg cleaners, candling

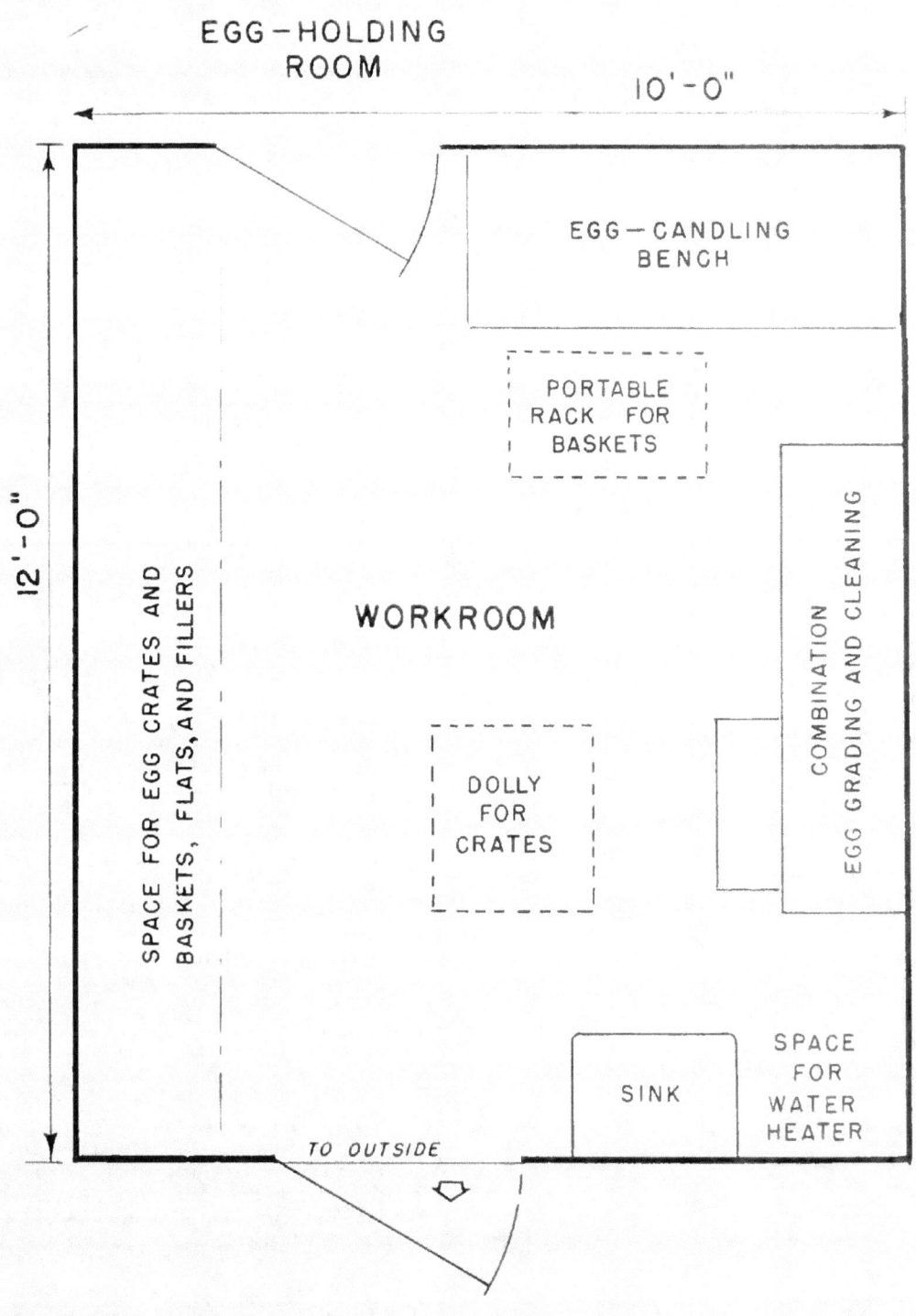

Figure 36.—Egg-handling and egg-holding room for a flock of 1,000 to 5,000 hens.

benches, and graders depends on the daily work routine. Ample space is needed to move filled egg crates and baskets. Hand trucks, dollies, and pallets save time and labor. A water heater may be desirable on a farm where eggs are washed.

Convenient heights for tables on which to pack egg crates while standing are 27 inches for a 6-foot person, and 24 inches for one 5½ feet tall.

Egg-holding room or cabinet

As eggs deteriorate rapidly in warm, dry, or odorous places, it is important that they be kept in a well-planned holding room or cabinet.

Eggs that are to be held should be cooled to 55° to 60° F. in less than 6 hours. Eggs for hatching should be stored at 55°. Relative humidity of the room or cabinet should be 75 to 85 percent.

For safe handling, eggs should be kept in baskets filled two-thirds to three-fourths. The baskets should either be hung, or be placed on slatted shelves, on racks with casters (fig. 37), or on fixed racks.

The size of the holding room or cabinet depends on the size of flock and the frequency of delivery of eggs to market. In most places, eggs are sold twice weekly; enough holding space for 4 days' eggs should therefore be provided. Prefabricated egg holding rooms or cabinets of suitable sizes complete with proper controls and refrigeration equipment are available.

The nearer the room or cabinet is to cubical shape, the less surface is exposed to heat, and the less building material is needed. As filled egg crates are not usually stacked more than 5 high, the minimum ceiling height should be 7 feet for the walk-in cooler, while cabinet coolers holding 15 cases of eggs or less may be 5 to 6 feet high.

The floor plans for holding rooms of various sizes are shown (fig. 38), together with the approximate size of refrigeration unit for each size of room. These plans are designed for farms that sell eggs twice weekly.

For simplest holding-room construction, 2- by 4-inch studs, spaced 24 inches apart, may be used for framing. Inside and outside walls and ceiling may be finished with pressboard, exterior grade plywood, or other moisture-resistant material. The space between the "2 by 4's" should be filled with insulating material, such as mineral wool, fiberglass blanket, or other insulation that does not settle. Blanket insulation is usually manufactured with vapor barrier attached to one side.

The most satisfactory floor for the holding room or cabinet is made of 3- to 4-inch-thick concrete, sloped ⅛ inch per foot to a drain. This floor, when laid over ground, should have a minimum of 3-inch gravel fill below it. If it is laid over an existing floor that is exposed to outdoor-air temperature, the floor of the holding room or cabinet should have the same amount of insulation as the walls or ceiling.

The doors of most holding rooms are sources of considerable heat gain because of thin, poor construction. These doors should be as thick and as well insulated as the walls and ceiling. It is important to install a safety latch so that the door can be opened from the inside.

When dollies or hand trucks are used, a threshold is not provided on the doorframe. Instead, a strip of reinforced rubber, ½ inch thick by 2 inches wide, is attached to the bottom of the door, so that heavy, cool air from the box does not leak out.

Designs for satisfactory door construction are shown (fig. 39).

As eggs are precooled in the baskets before packing, various methods can be used to avoid warming the holding room when the main door is opened. One method is to insert a small opening in the wall at waist height, through which egg crates and baskets may be moved.

If the holding room is painted, odorless paint, which is especially manufactured for rooms containing fresh foodstuffs, should be used. This is important, because all odorous materials should be kept out of egg-holding rooms to prevent eggs from taking on the odors.

Nearly all refrigeration units on poultry farms are air-cooled. In appearance, these units resemble window type air conditioners but they are especially designed for egg-holding rooms or cabinets. Refrigeration-blast coils may be used (fig. 38).

The fans on the cooling coils should run continually. These coils should operate without frosting, to maintain a room temperature of 50° to 65° F. Relative humidity of the room should be 75 to 85 percent. Automatic humidifiers that have thermostatic control should be used, to maintain this level of humidity with the correct room temperature.

For economical operation of the compressor unit, it is essential to keep it clean and, when possible, to install it in a room where the air is free from dust.

Electrical outlets, lights, and standby generators

During fall and winter, or on cloudy or foggy days, artificial lights are desirable to maintain

FIGURE 37.—Portable racks for egg baskets in holding room.

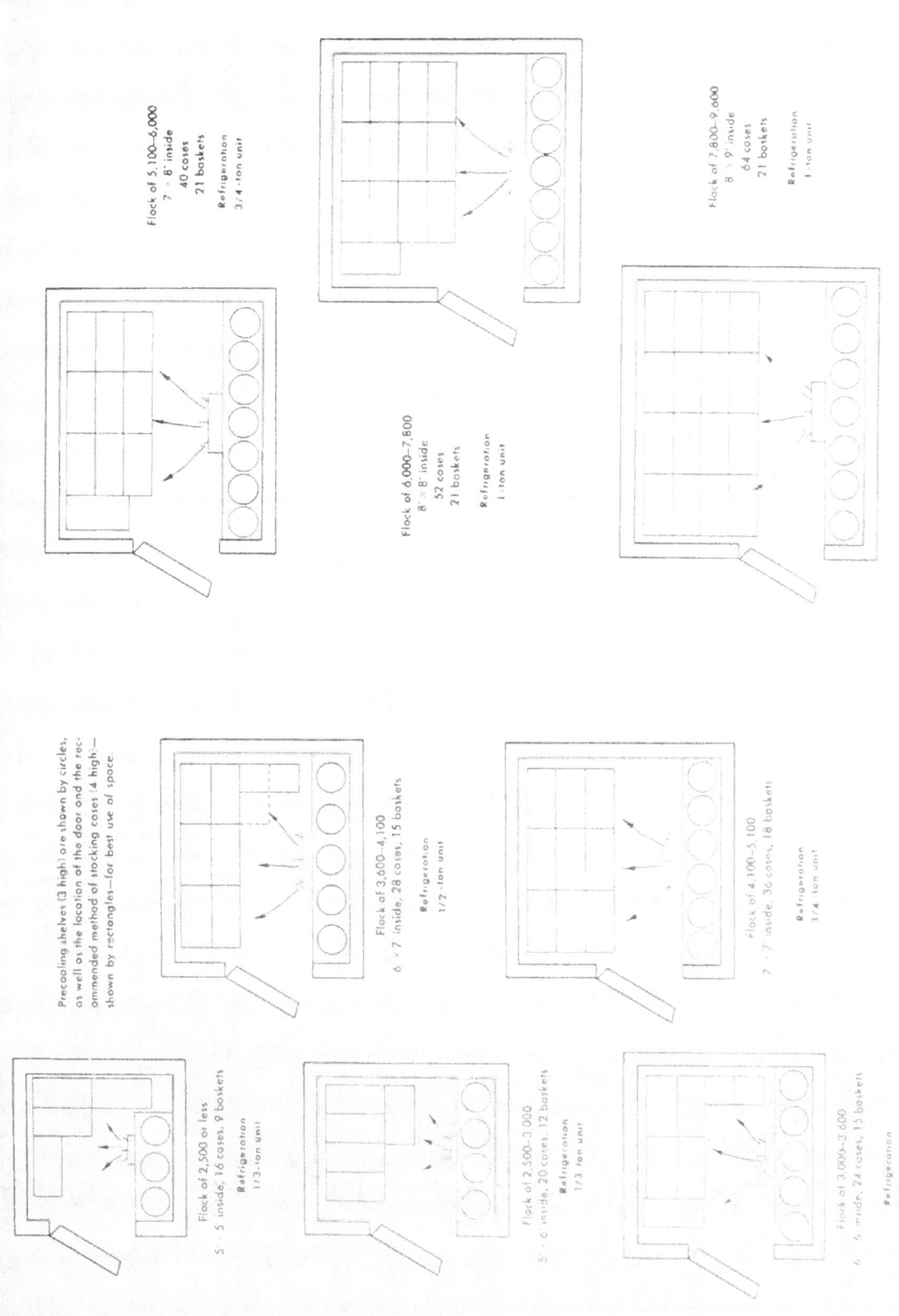

Figure 28. Floor plans for holding rooms of various sizes. Each has shelves for egg baskets, stacked egg crates, and blast cooling coil. The approximate size of refrigeration unit for each room is given in tons.

Source: University of California

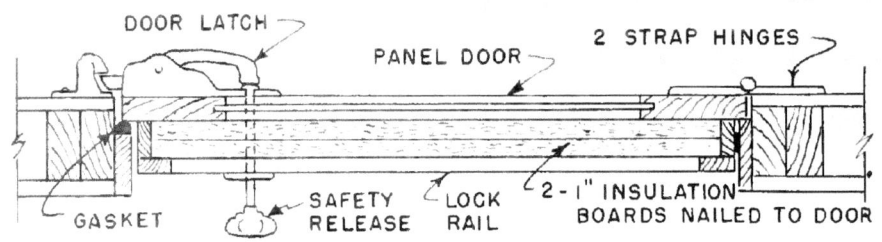
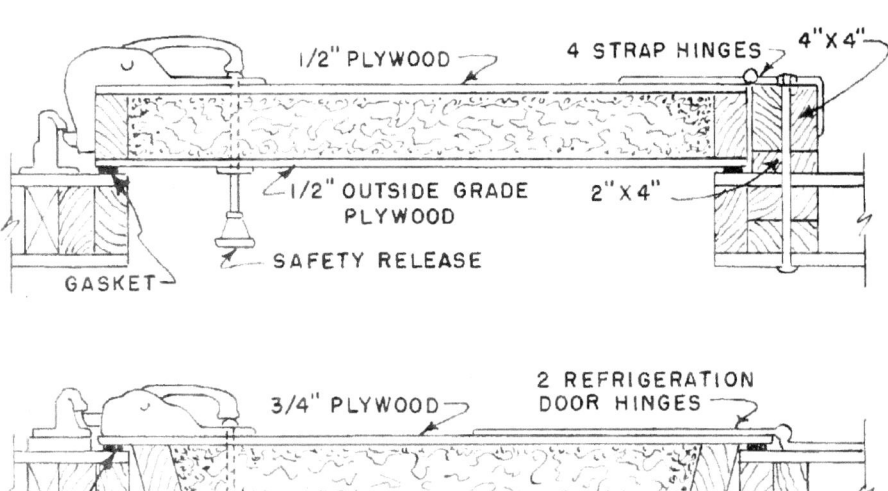
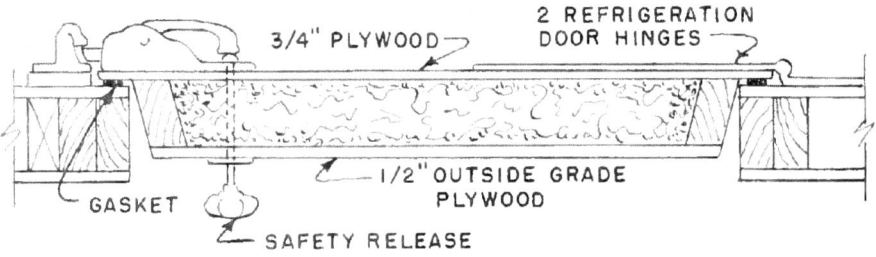

FIGURE 39.—Satisfactory door designs with safety latches.

egg production. A total daily light period of 13 to 14 hours is sufficient. The preferred method is to turn the lights on by an automatic time clock early in the morning. There should be one 40- to 60-watt incandescent lamp or one 15-watt fluorescent lamp for each 200 square feet of floor area. These lamps should be spaced 10 feet apart and installed 5 feet away from walls or solid partitions. The lights should be adjusted to illuminate the entire floor and roosting areas.

All-night lighting should provide just enough light to keep the hens "working" late in the evening and to start their day earlier. For night lighting, a 10- to 15-watt incandescent lamp is needed for each 200 square feet. This lamp should be placed above feeding and watering area.

In both the egg-handling and the egg-holding room, enough lights should be used to minimize shadow.

Outdoor lights and burglar alarms are desirable around poultry houses to discourage robbery. These may be mounted on poles or on nearby buildings, and may be turned on or set from the dwelling or operated by automatic controls.

Where water-warming devices are used, 1 outlet is needed for each 400 square feet, or at least 1 outlet for each pen.

All wiring should be installed by qualified personnel and should be checked periodically for safety.

To prevent severe slumps in egg production, caused by light outage during power failure, a standby generator is desirable to take care of minimum requirements for lighting and pumping water for large flocks. The generator should be located in a small building by itself. Local power companies are prepared to give advice regarding the selection of equipment.

The approximate power requirements for various items of equipment are as follows:

Equipment:	Horsepower
Automatic poultry feeder	$\frac{1}{4}-\frac{3}{4}$
Portable elevator	$\frac{1}{2}-1\frac{1}{2}$
Platform hoist	$\frac{1}{2}-1$
Litter stirrer	$\frac{1}{4}-\frac{1}{2}$

For small flocks, lanterns may be used for emergency lighting.

www.ingramcontent.com/pod-product-compliance
Lightning Source LLC
Chambersburg PA
CBHW062207220526
45470CB00009B/2951